MECHANICAL ENGINEERING THEORY AND APPLICATIONS

LIGHT MACHINE TOOLS FOR PRODUCTIVE MACHINING

MECHANICAL ENGINEERING THEORY AND APPLICATIONS

Additional books in this series can be found on Nova's website under the Series tab.

Additional E-books in this series can be found on Nova's website under the E-book tab.

Mechanical Engineering Theory and Applications

Light Machine Tools for Productive Machining

J. ZULAIKA
F. J. CAMPA
AND
L.N. LOPEZ LACALLE

Nova Science Publishers, Inc.
New York

Copyright © 2011 by Nova Science Publishers, Inc.

All rights reserved. No part of this book may be reproduced, stored in a retrieval system or transmitted in any form or by any means: electronic, electrostatic, magnetic, tape, mechanical photocopying, recording or otherwise without the written permission of the Publisher.

For permission to use material from this book please contact us:
Telephone 631-231-7269; Fax 631-231-8175
Web Site: http://www.novapublishers.com

NOTICE TO THE READER

The Publisher has taken reasonable care in the preparation of this book, but makes no expressed or implied warranty of any kind and assumes no responsibility for any errors or omissions. No liability is assumed for incidental or consequential damages in connection with or arising out of information contained in this book. The Publisher shall not be liable for any special, consequential, or exemplary damages resulting, in whole or in part, from the readers' use of, or reliance upon, this material. Any parts of this book based on government reports are so indicated and copyright is claimed for those parts to the extent applicable to compilations of such works.

Independent verification should be sought for any data, advice or recommendations contained in this book. In addition, no responsibility is assumed by the publisher for any injury and/or damage to persons or property arising from any methods, products, instructions, ideas or otherwise contained in this publication.

This publication is designed to provide accurate and authoritative information with regard to the subject matter covered herein. It is sold with the clear understanding that the Publisher is not engaged in rendering legal or any other professional services. If legal or any other expert assistance is required, the services of a competent person should be sought. FROM A DECLARATION OF PARTICIPANTS JOINTLY ADOPTED BY A COMMITTEE OF THE AMERICAN BAR ASSOCIATION AND A COMMITTEE OF PUBLISHERS.

Additional color graphics may be available in the e-book version of this book.

LIBRARY OF CONGRESS CATALOGING-IN-PUBLICATION DATA

Zulaika, J.
 Light machine tools for productive machining / authors, J. Zulaika, F.J. Campa, and L.N. Lopez Lacalle.
     p. cm.
 Includes index.
  ISBN 978-1-61324-644-3 (softcover)
  1. Machine-tools. I. Campa, F. J. II. Lspez de Lacalle, L. N. III. Title.
  TJ1185.Z78 2011
  621.9'02--dc23

2011017310

Published by Nova Science Publishers, Inc. ✛ New York

CONTENTS

Chapter 1	Introduction	1
Chapter 2	Design Requirements for Achieving Specific Productivity Levels	5
Chapter 3	Application of the Approach on an Actual Industrial Case	27
Chapter 4	Validation of the Prototype of Productive and Eco-Efficient Milling Machine	41
Chapter 5	Conclusion	47
References		49
Index		51

Chapter 1

INTRODUCTION

The *eco-efficiency* of a machine tool can be defined as the ratio between the extent to which its functional objectives have been met combined with the environmental impact associated with its total lifecycle. In quantitative terms, machine eco-efficiency can be expressed as the ratio between the productivity of the machine and the environmental impact associated with its lifecycle.

With the aim of measuring the *environmental impact* associated with the lifecycle of a machine tool, the authors carried out a *Life Cycle Assessment* (LCA) on a milling machine made by "Nicolas-Correa" Company, considering a life span of 10 years for the machine. When conducting the LCA, the authors followed ISO 14040 rules and selected the "Climate change" impact category within the Ecoindicator 99 methodology [1]. The main conclusion that the authors reached from the LCA was that much of the environmental impact of a machine tool – around 95% - was associated with its use phase. Furthermore, within the use phase, 95% of its impact was produced by the energy that the machine consumed. Figure 1 shows these data both graphically and numerically.

The *productivity* associated with a machine tool or with a machining process is usually measured in terms of *Material Removal Rate* (MRR), which is defined as the volume of material removed per unit of time. The MRR, usually referred to as Z_w, is calculated as the product between the cutting area and the feed velocity of the tool. In the great majority of cases, this productivity is limited by the appearance of self-induced vibrations, vibrations that tend to be commonly referred to as *chatter*. Among the types of self-excited vibrations, attention should be paid to *regenerative chatter* [2], due to both the frequency with which it occurs and its intensity. This is the self-excited vibration *par excellence* in the machining process, and is caused by the regeneration of the metal thickness in systems in which the blade cuts a previously machined surface either totally or partially [3]. This phenomenon will be analysed in detail in next paragraphs.

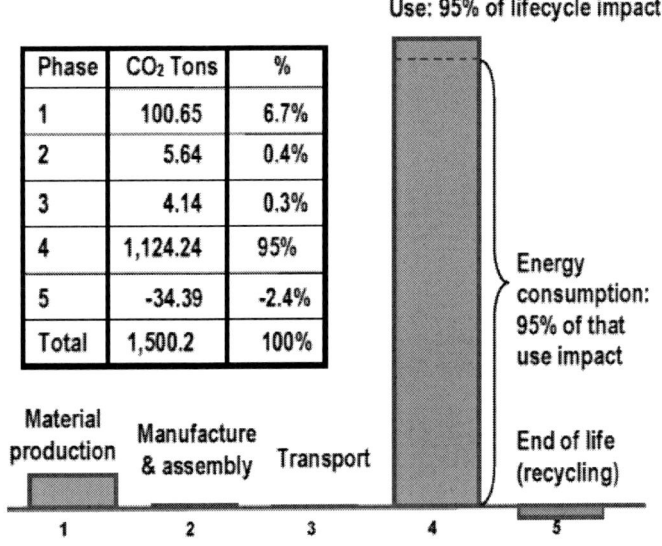

Figure 1. Life Cycle Assessment Study on a milling machine [1].

This work will integrate the two concepts of eco-efficiency and productivity, that traditionally have been viewed as being mutually exclusive goals. Currently, standard practice for machine tool builders is to produce high-performance machines by means of high dynamic feed-drives [4], which forces design engineers to utilize heavy and stiff mechanical structures capable of bearing the higher inertial forces. The increased weight of structural elements in turn creates a requirement for more powerful actuators, leading to higher inertial forces. This is a vicious circle in which the increase of productivity in machining processes has been achieved at the expense of high material and energy consuming machines, increasing both the environmental impacts and costs associated with machining processes.

Towards a goal of building machine tools that are both highly productive and eco-efficient, this book presents an integrated 'machine and process' approach that implements separating the productivity of machining processes from the material intensity of machine tools. The details of this approach are based on achieving a specific value of productivity in machining processes by constructing machines using the minimum amount of material in their structural elements, leading to a reduction in both the material resources consumed in the production phase and in the energy resources consumed by the machines during their use phase.

The conceptual cornerstone for this machine *dematerialization* process lies in defining the exact dynamic and mechanical performances that are required for the machine to assure specific productivity goals, and achieving them with the least possible weight of mobile structural elements.

At this point, it is worthwhile to reflect on the functionality of the structural elements of a machine tool. On the whole, the structure of a machine tool has two main functions [5]:

1) To hold the components and peripherals involved in the machine and to assure their kinematic requirements during machine motions and processes.
2) To withstand the forces that are produced by the machining process and by the machine motions in a robust manner for assuring the required motion accuracy.

Concerning these two functions, it is feasible to fulfil the former requirement - i.e. to hold machine components and provide the required kinematic performances - with structural components of much lower weight. An example of this is the case of industrial robots, which are much lighter than machine tools. In quantitative terms, according to a study of the authors, it is possible to assure the kinematic and dynamic performances of machine components with approximately 25% of the weight of the mechanical structure of an average milling tool [6]. This means that the functionality of the remaining 75% of the mechanical structure is only to assure robust performance of the machine during machine positioning and production processes. Therefore, for a specific machining process, if the specifics of the minimum requirements for stiffness and robustness are identified and quantified, and they are achieved using a minimum of structural weight, the result is a wide range of possibilities for reducing the weight of a machine in a significant manner without affecting its productivity.

Following this conceptual approach, the next chapters will analyse on the one hand the mechanical design requirements associated with specific values of process productivity and accuracy, and on the other hand, will analyse synergetic and complementary approaches for assuring process productivity while using a minimum of material in the machine structure.

Chapter 2

DESIGN REQUIREMENTS FOR ACHIEVING SPECIFIC PRODUCTIVITY LEVELS

With the aim of defining the design requirements that lead to productive and stable machining processes, the mechanical requirements that are associated with both static and dynamic stiffness of the machine tool structures will be analysed.

2.1. REQUIREMENTS OF STATIC STIFFNESS FOR PRODUCTIVE AND ACCURATE MACHINE TOOLS

Table 1 shows the average process forces associated with milling operations when cutting AISI 1045 Steel [7]. These data have been measured experimentally by the authors on various machines in different workshops, and can be used as basis for defining threshold values for static stiffness of a machine tool at the *tool centre point* (TCP) that will assure an accurate and reliable performance of the machine. These results have been combined with accuracy requirements that machine end-users have provided for the different milling operations, which has allowed the authors to develop operational specifications associated with different machining operations.

Table 1. Average forces and acceptable deformations for different milling operations

Process	Average max. force on TCP	Acceptable deformation
Roughing	Conventional tools: 1,500 N 100-125 mm. diam. tools: 3,000 N	Average: 100 µm
Semi-finishing	Conventional tools: 1,000 N	Average: 50 µm
Finishing	Conventional tools: 200 N	Average: 10 µm

Combining the actual forces that have been gathered in the experimental measurements with the precision requirements that the end users have provided to the authors, and taking into account that the main component of machining forces is in the tangential direction of the tool, a static stiffness of 1500N/100µm (i.e. 20 N/µm) at the tool centre point in the tangential direction can be considered as a reference threshold value for general-purpose milling machines. This means that if a machine has a higher static stiffness, from the static point of view there is margin for reducing the weight of the machine without affecting the accuracy and reliability of machining processes.

2.2. REQUIREMENTS OF DYNAMIC STIFFNESS FOR PRODUCTIVE AND RELIABLE MACHINE TOOLS

In addition to the requirements associated with the static stiffness of a machine tool, it is worthwhile to highlight that in the great majority of machining processes, the productivity of machining processes is not limited by static deformations but by the presence of vibrations, in particular of self-induced vibrations. Among these self-induced vibrations, special attention should be paid to *regenerative chatter,* owing both to its incidence and to its intensity. This regenerative chatter is caused by the regeneration of the metal thickness in systems in which the blade cuts a previously machined surface either totally or partially, and its appearance is linked to the following parameters [8, 9]:

I. The dynamic stiffness of the machine-process system;
II. The geometry of the machining tool;
III. The radial immersion of the machining tool into the material to be removed
IV. The material to be removed.

From this perspective, if the part to be machined is of a hard material such as steel, and its dimensions are below the dimensions of the machine, the dynamic performance of the system will be dominated by the natural frequencies of the machine structure, as shown in Figure 2 [10].

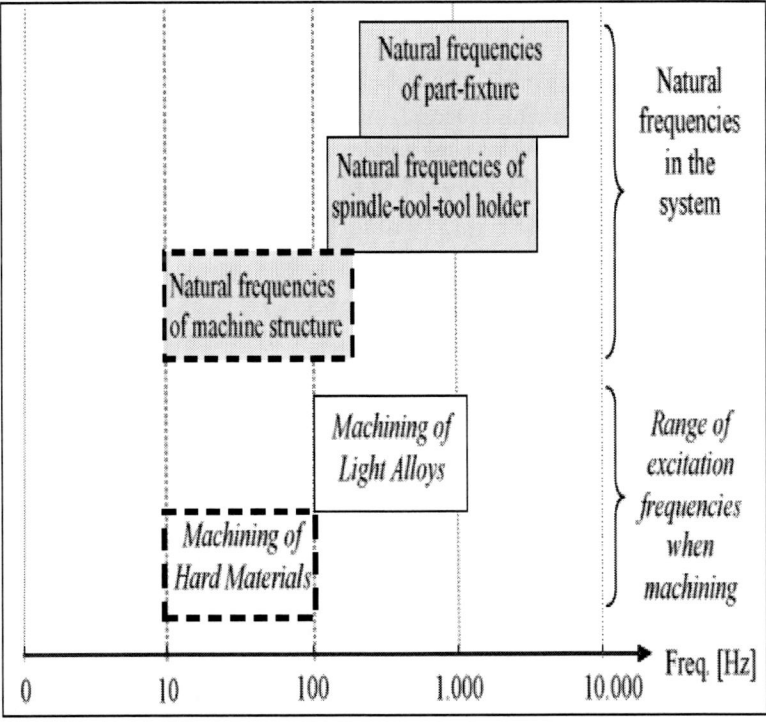

Figure 2. Ranges for natural frequencies of the system *(upper part, grey-shaded boxes)* and for excitation frequencies when machining different materials *(blank boxes)* [10].

As can be seen in Figure 2, the natural frequencies of the machine-process system at low frequencies (below 200 Hz.) are related to structural natural frequencies of the machines, whereas the natural frequencies above

200 Hz. are related to natural frequencies of machine components such as spindle, tool, tool-holder etc. The manufacturers distinguish between two general types of chatter, the so-called 'structural chatter' or 'machine chatter', which is a low frequency chatter, and the so-called 'tool chatter', which appears at higher frequencies. In the former case, recognisable as a low-pitched sound, the chatter is associated with the structural modes of the machine, whereas in the latter case, marked by a high-pitched sound, the chatter is associated with the modes of specific machine components such as spindle, tool etc.

The type of chatter that will be excited by the machining forces will depend on the excitation frequencies associated with those machining processes. According to Figure 2, when machining hard materials such as steel, the associated excitation frequencies will be in the range of the natural frequencies of the structural elements of the machine. Therefore, the stability of this type of machining processes will be affected mainly by the modal parameters associated with the *machine structure*.

Within this view, in the next subchapter a stability model adapted to the machining of parts of hard materials that are smaller than the machine dimensions will be developed, which will serve as basis for translating process stability requirements into design requirements in terms of dynamic stiffness at the TCP.

2.2.1. Stability Model for the Machining Process

As starting point, a stability model has been developed for milling processes that is based on a mechanistic modelling of the milling forces [8], a mono-frequency approach of the stability solution [11], the introduction of the dynamics of the machine tool in modal coordinates as in [12], and the influence of the feed direction as in [13]. In addition, the modal vector at the tool tip is referenced to the machine tool axes X_{MT} Y_{MT} Z_{MT} and to the local axes of the tool X_t Y_t Z_t, coinciding Z_{MT} and Z_t axes and defining X_{MT} and Y_{MT} the working plane of the machine and $+X_T$ the feed direction of the tool, as shown in Figure 3. $\bar{i}_m$ is the versor that defines the direction of a mode and $\{\phi_i\}$ the modal vector per mass unity, the projection of the versor over the $X_{MT}Y_{MT}$ plane is $\bar{i}_m^{xy}$, whose angle with respect to the feed direction is β_{xy}, whereas the angle between the modal vector and the $X_{MT}Y_{MT}$ plane is β_z. Hence, the direction of a given machine mode is related to the feed direction $(+X_T)$ through the two angles β_{xy} and β_z.

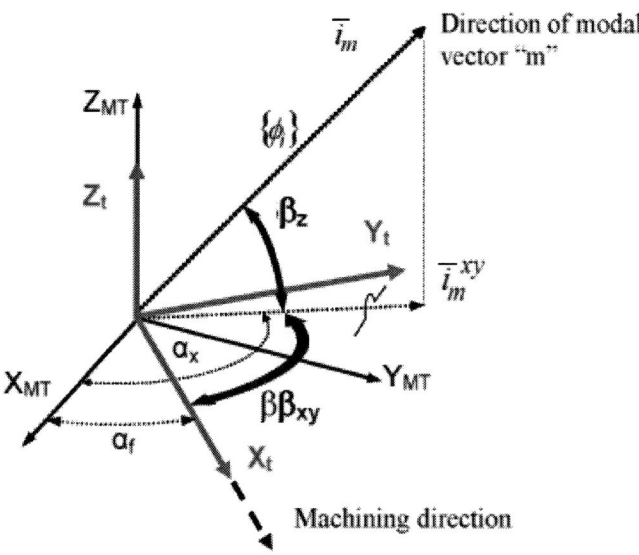

Figure 3. Orientation of a machine modal vector in machine tool axes *(MT indexes)* and tool axes *(t indexes)*.

For a jth cutting edge of a milling tool, the mechanistic cutting forces model proposes a linear relation between the dynamic chip thickness h_d and the width of cut b in the cutting area (see Figure 4 below right) with the tangential, radial and axial forces F_t, F_r and F_a (see Figure 4 below) that are acting over the cutting edge, being the linear relation based on tangential, radial and axial *cutting coefficients* K_{tc}, K_{rc} and K_{ac} respectively:

$$F_{tra,j}(\phi_j) = \begin{Bmatrix} F_{t,j} \\ F_{r,j} \\ F_{a,j} \end{Bmatrix} = \begin{Bmatrix} K_{tc} \\ K_{rc} \\ K_{ac} \end{Bmatrix} \cdot b \cdot h_d(\phi_j) = \frac{K_t \cdot a_p}{\sin \kappa} \begin{Bmatrix} 1 \\ K_r \\ K_a \end{Bmatrix} h_d(\phi_j), \quad (1)$$

where ϕ_j is the angular position of that jth cutting edge related to the $+Y_t$ axis (see Figure 4 below left), and the width of cut b can be related to the axial depth of cut a_p by means of the cutting edge lead angle of the insert κ, $b = a_p/\sin\kappa$ (see Figure 4 below right). In addition, for the sake of clarity, the cutting coefficients K_{tc}, K_{rc} and K_{ac} can be normalised with respect to the tangential component $K_t = K_{tc}$, as shown in equation (1), being $K_r = K_{rc}/K_{tc}$ and $K_a = K_{ac}/K_{tc}$.

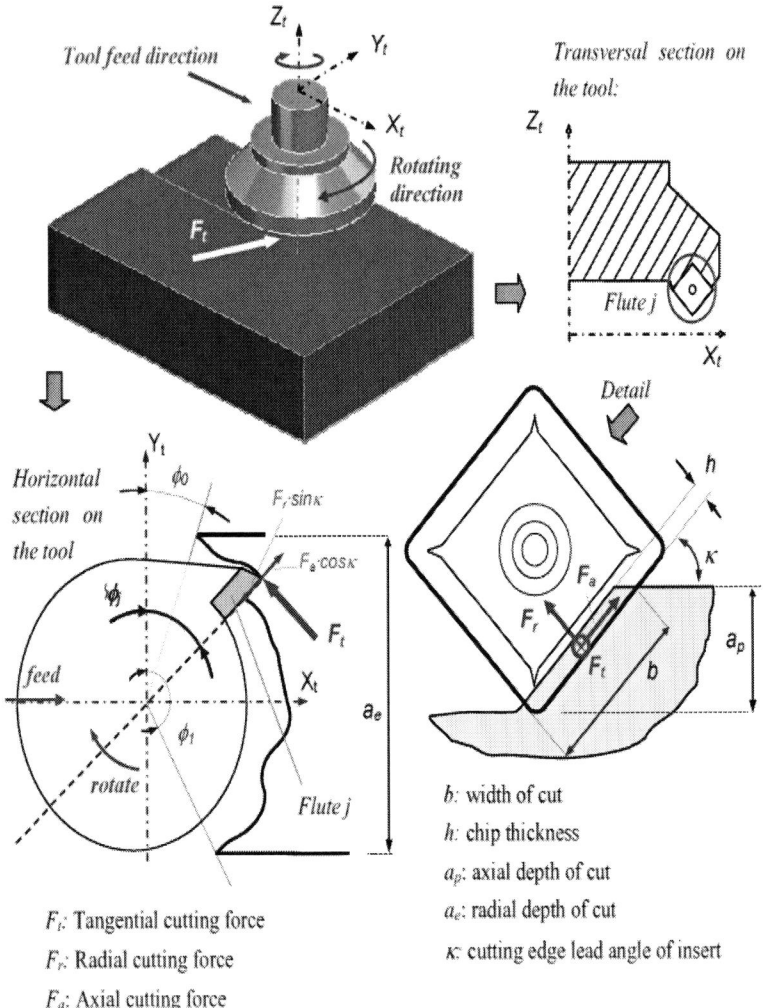

Figure 4. Cutting forces involved in the machining process.

By means of this mechanistic model for the machining forces and a dynamic model of the machine, the dynamic machining forces that are acting on "j" flute can be projected over the mth mode direction $\overline{i_m}$, calling a_F^j to those projections. Bearing in mind the modal vector modulus ϕ_i for the ith mode and considering n modes of the machine, the process forces can be expressed in modal coordinates g_m:

$$\begin{Bmatrix} g_{1,j} \\ g_{2,j} \\ \vdots \\ g_{n,j} \end{Bmatrix} = [\{\phi_1\} \; \{\phi_2\} \; \cdots \; \{\phi_n\}]^T \cdot \begin{bmatrix} a_{F_t}^1 & a_{F_r}^1 & a_{F_a}^1 \\ a_{F_t}^2 & a_{F_r}^2 & a_{F_r}^2 \\ \vdots & \vdots & \vdots \\ a_{F_t}^n & a_{F_r}^n & a_{F_r}^n \end{bmatrix} \cdot F_{tra,\,j} \, . \quad (2)$$

On the other hand, each of these "n" modes has an associated modal displacement in the mode direction that will be called δ. In this respect, naming as Δm_i the dynamic deformation in global coordinates that is associated with a mode "i" of the machine at the tool centre point, that displacement will be expressed as a modal displacement $\Delta \delta_i$ by means of the following expression, where $\{\phi_i\}$ is the modal vector per mass unity for that mode "i" at the tool centre point:

$$\Delta m_i = \phi_i \Delta \delta_i \, . \quad (3)$$

Generally, the change from global coordinates y_i into modal coordinates δ_i is conducted by means of the following transformation in the Laplace domain:

$$\{\delta(s)\} = [\Phi]\{y(s)\}. \quad (4)$$

where $[\Phi]$ represents the matrix of modes, whose columns are the "n" eigen-modes of the considered mechanical system:

$$[\Phi] = [\{\phi\}_1 \, \{\phi\}_2 \, \cdots \, \{\phi\}_j \, \cdots \, \{\phi\}_n]. \quad (5)$$

Thus, in a machine where "n" modes have been considered, the effect of each of the modal displacements $\Delta \delta_i$ h_d on the dynamic chip thickness will become:

$$h_d(\phi_j) = \{\phi_1 a_{h_1} \; \phi_2 a_{h_2} \; \cdots \; \phi_n a_{h_n}\} \begin{Bmatrix} \Delta \delta_1 \\ \Delta \delta_2 \\ \vdots \\ \Delta \delta_n \end{Bmatrix}. \quad (6)$$

where the terms a_{h_i} represent the projection of the modal displacements in the direction of the chip thickness h_d, as shown in Figure 5, and takes the following expression:

$$a_{h_i} = (sin\phi_j cos\beta_{xy} + cos\phi_j sin\beta_{xy})cos\beta_z \cdot sin\kappa - cos\kappa \cdot sin\beta_z . \qquad (7)$$

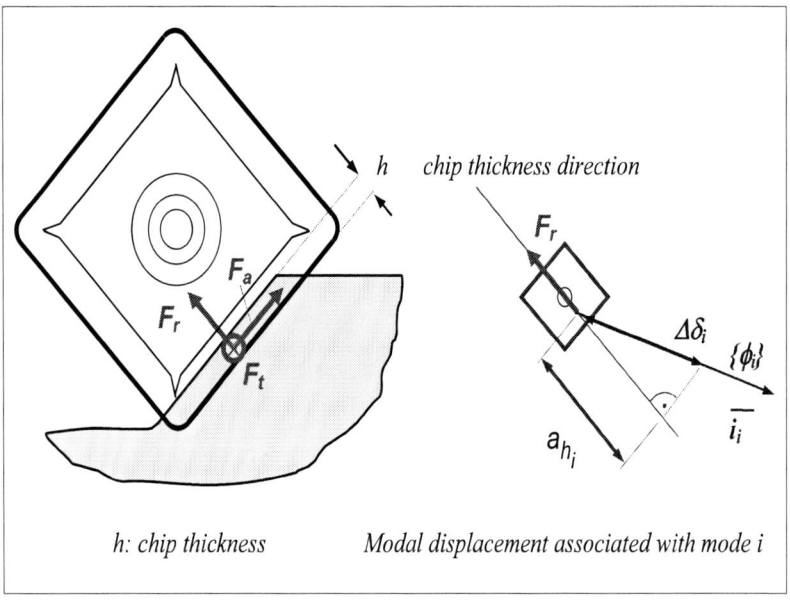

Figure 5. Effect of modal displacements on the dynamic chip thickness h_d.

Now that both forces and displacements have been expressed in modal coordinates, integrating the former equations, the following relation is achieved with respect to modal forces and displacements in a machining process [14]:

$$g = K_t \cdot a_p \cdot A(\phi) \, \Delta\delta/2, \qquad (8)$$

where K_t represents the tangential cutting coefficient, a_p the axial depth of cut, $\Delta\delta$ the dynamic displacement of the tool in modal coordinates and $A(\phi)$ a matrix of dimensionless factors.

Design Requirements for Achieving Specific Productivity Levels 13

Matrix $A(\phi)$ consists of dimensionless factors called *directional factors*, where for the row "p" and the column "q", the element a_{pq} of the matrix takes the following expression:

$$a_{pq} = \frac{2 \cdot \phi_q \cdot a_{h_q} \cdot \phi_p}{\sin \kappa} \left[a_{F_t}^p + K_r \cdot a_{F_r}^p + K_a \cdot a_{F_a}^p \right]. \tag{9}$$

According to equation (2), a term a_F^i represents the projection of a force F on the direction of a mode "i", whereas according to equation (7), a_{h_i} represents the projection of a modal displacement that is along the direction of a mode "i" on the direction of the dynamic chip thickness h. Thus, a a_{pq} element of that matrix indicates to which extent a dynamic displacement along the direction of a mode "p" affects the chip thickness h and by extension the generation of process forces that excite the mode "q". This coefficient can be either positive or negative; a positive coefficient means that the machining forces make the tool separate from the workpiece, whereas a negative coefficient means that the machining forces make the tool penetrate the workpiece.

When the cutting forces excite any of the structural modes of the machine that is holding the tool, a wavy surface finish is left due to structural vibrations. That wavy surface left by the previous tooth is removed during the succeeding tooth period, which also leaves a wavy surface due to the dynamic cutting forces [15]. Thus a self-excitation mechanism is generated during the material removal process, in which the output process variable – the position of the tool- affects the input process variable –the chip thickness-, generating a positive feedback system.

On the other hand, depending on the phase shift between two successive waves, the chip thickness may grow exponentially while oscillating at a chatter frequency ω_c that will be close to the dominant structural mode of the machine, i.e. the natural frequency of the open loop system in Figure 6.

Based on the block diagram that is shown in Figure 6, a mono-frequency solution is applied on it to solve its characteristic equation. Thus, if chatter appears at a frequency ω_c, the dynamic displacement $\Delta \delta_i$ of an *i*th mode in modal coordinates and for a tooth passing period τ can be obtained from the dynamic cutting forces and from its modal parameters. Namely, the involved parameters are its natural frequency ω_i, its modal vector ϕ_i, its damping coefficient ξ_i and its effective stiffness $k_{efi} = \omega_i^2 / \phi_i^2$, and the displacement takes the following expression [14]:

$$\Delta\delta_i = (1-e^{-i\omega_c\tau})\cdot h_i(\omega)\cdot g_i, \qquad (10)$$

in which h_i is the transfer function of the machine at the TCP in modal coordinates. If the variable $r_i=\omega_c/\omega_i$ is defined, the transfer function h_i in modal coordinates takes the following form:

$$h_i(\omega) = \frac{1/\omega_i^2}{(1-r_i^2)+i\cdot(2\xi_i r_i)^2}. \qquad r_i=\omega_c/\omega_1 \qquad (11)$$

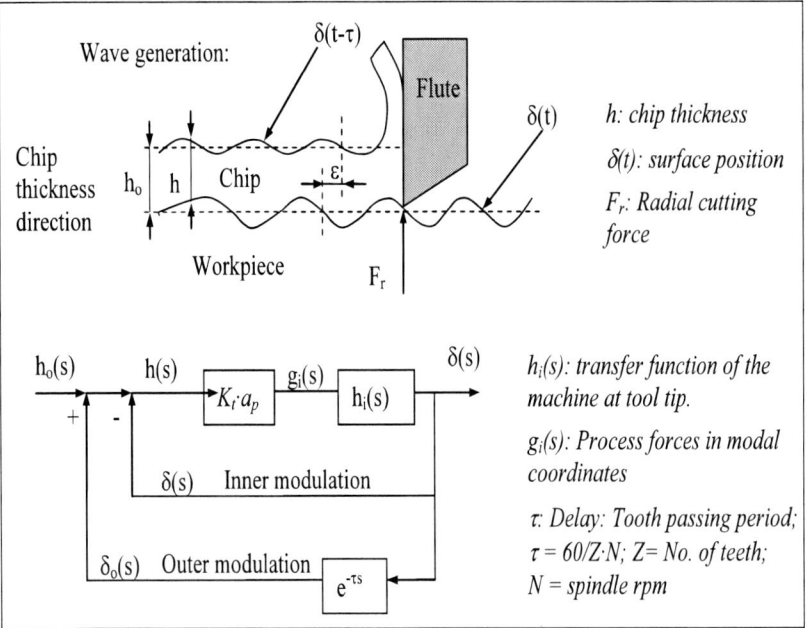

Figure 6. Block diagram of chatter dynamics [15].

Thus, for the case of a machine with "n" structural eigen-modes, the total dynamic displacements of the "n" modes will take the following form:

$$\Delta\delta = (1-e^{-i\omega_c\tau})\cdot H(\omega)\cdot g, \qquad (12)$$

in which $H(\omega)$ is a diagonal matrix nxn in which the elements of the diagonal are the transfer functions h_i in modal coordinates:

$$H(\omega) = \begin{bmatrix} h_1(\omega) & 0 & \cdots & 0 \\ 0 & h_2(\omega) & \cdots & 0 \\ \vdots & \vdots & \ddots & \vdots \\ 0 & 0 & \cdots & h_n(\omega) \end{bmatrix}. \tag{13}$$

If the previous equations are combined and the mean Fourier term of the directional coefficient matrix is adopted, naming α the resulting matrix, the *eigen-equation* (14) is reached, in which Z is the number of flutes on the tool:

$$g = \frac{Z}{4\pi} \cdot K_t \cdot a_p \cdot (1 - e^{-i\omega_c \tau}) \cdot \alpha \cdot H(\omega) \cdot g. \tag{14}$$

2.2.2. Stability Problem Solution

The solution of the *nth* order eigenvalue problem in modal coordinates shown in (14) provides the relation between the chatter frequency, the critical depth of cut and the spindle speed, and allows assembling the stability lobes and chatter frequency diagrams, as in [11].

As the equations are expressed in modal coordinates, they are decoupled among them except for the case of the α matrix of directional factors, which is not symmetrical. If the asymmetry of the α matrix is neglected, each mode can be analysed in an individual manner.

Thus, for equation (14), for the specific case of considering *one concrete structural mode "q" of the machine* (actually, in the real machining cases, for one specific machine position and for one specific machining direction, there will always be *one* predominant eigen-mode of the machine that will cause process instabilities), and finding the depth of cut a_p in that expression, the resulting value is the following:

$$a_p = \frac{2\pi}{Z \cdot K_t} \cdot \frac{1}{\alpha} \cdot \frac{1}{real(h(\omega))}, \tag{15}$$

where α is the *directional coefficient* associated with the considered eigen-mode "q". Basing on equation (9), the average Fourier term for the directional factor associated with that mode "q" takes the following form:

$$\alpha_q = \frac{2}{\sin\kappa} \cdot \phi_q^2 \cdot \int_{\phi_0}^{\phi_1} a_{h_q} (a_{F_t}^q + K_r \cdot a_{F_r}^q + K_a \cdot a_{F_a}^q) d\phi . \tag{16}$$

Coming back to equation (15), for the specific case of a mechanical system of one degree of freedom, the real part of a transfer function with a mechanical stiffness k and a modal damping ξ takes the form that is shown in Figure 7.

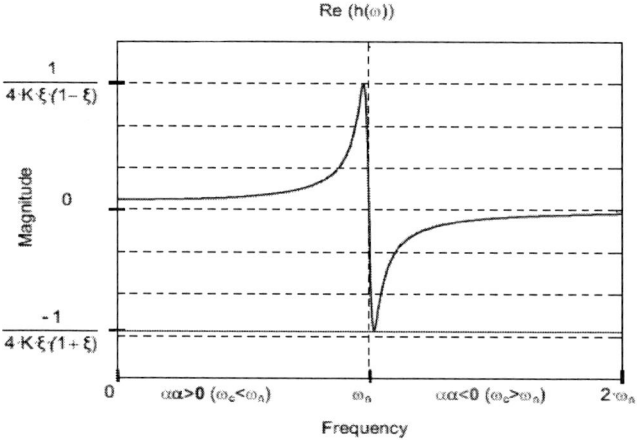

Figure 7. Real part of the transfer function of a system of one degree of freedom.

As can be seen in Figure 7, the maximum and the minimum values of the real part of the transfer function take the following form:

$$max/min(Re(h(\omega))) = \frac{1}{4 \cdot k \cdot \xi \cdot (1 \mp \xi)} . \tag{17}$$

On the other hand, a new term σ can be defined as the *real directional factor*, for removing the modal vector ϕ_q in the expression of the *directional factor* α_q:

$$\sigma = \frac{\sin\kappa}{2 \cdot \phi_q^2} \cdot \alpha_q . \tag{18}$$

Combining thus the equations (15) to (18) and finding the expression of the critical depth of cut a_{pcrit}, the following expression is obtained (19):

Design Requirements for Achieving Specific Productivity Levels 17

$$\sigma < 0 \rightarrow a_{P_{crit}} = \frac{4\pi \cdot \sin\kappa}{Z \cdot K_t} \cdot \frac{1}{\sigma} \cdot \frac{k_{ef}\xi(1+\xi)}{1} \cong \frac{4\pi \cdot \sin\kappa}{Z \cdot K_t} \cdot \frac{1}{\sigma} \cdot k_{ef} \cdot \xi$$

(19)

$$\sigma > 0 \rightarrow a_{P_{crit}} = \frac{4\pi \cdot \sin\kappa}{Z \cdot K_t} \cdot \frac{1}{\sigma} \cdot \frac{k_{ef}\xi(1-\xi)}{1} \cong \frac{4\pi \cdot \sin\kappa}{Z \cdot K_t} \cdot \frac{1}{\sigma} \cdot k_{ef} \cdot \xi$$

k_{ef} is the effective stiffness of the considered mode ($k_{ef} = \omega_n^2/\phi^2$, being ϕ the modal vector per mass unit), that in a system of one degree of freedom its value coincides with the value of the static stiffness k, $k_{ef} = k$. Furthermore, κ is the lead angle of the cutting edge and σ is the real directional factor, a dimensionless factor that depends on the orientation between the mode and the feed direction as well as on the normalized radial and axial cutting coefficients [13].

Figure 8 shows a 2D map of the values that this coefficient can reach with respect to β_{xy} and β_z angles. If $|\sigma|<1$ this coefficient increases the value of the modal effective stiffness k_{ef}, and in turn if $|\sigma|>1$ it decreases that value.

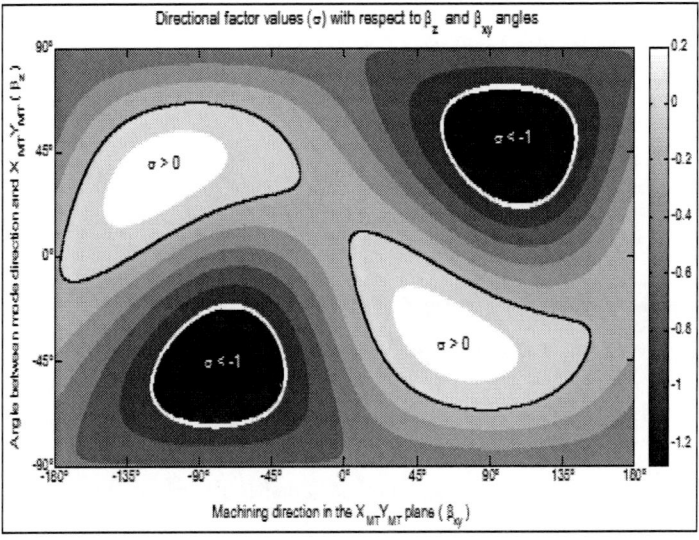

Figure 8. 2D Map of values of the directional factor σ with respect to β_{xy} and β_z angles.

Finally, it must be noted that in equation (19), an approximation of $(1+\xi) \approx 1$ has been applied, since the damping coefficient ξ that is associated with structural models of machine tools is negligible compared with a value of unity. Indeed, the authors have measured an average value of $\xi=0.02$ for structural modes of several milling machines of different architectures.

2.2.3. Requirements of Dynamic Stiffness for a Specific Productivity Level

As the productivity of a machining process is directly proportional to the depth of cut a_p, according to (19), to obtain a specific level of productivity in a machine tool in terms of MRR, the specific structural mode of the machine that limits the stability of the machining process requires a threshold value for the product of the *effective stiffness* k_{ef} and the *damping coefficient* ξ associated with that limiting mode. It is also notable that the real directional factor σ may increase or decrease the actual value of the modal effective stiffness.

Thus, coming back to (19), considering the machine tool as a system of one degree of freedom with associated effective stiffness k_{ef} and damping factor ξ, and defining a factor V as V=((Z·K$_t$)/(4·πsinκ)), the *dynamic requirement* associated with the *productivity of the machining process, per millimetre of depth of cut* a_p, turns out to be:

$$\frac{1}{\sigma} \cdot k_{ef} \cdot \xi \geq V \ . \qquad (V=Z \cdot K_t / 4 \cdot \pi sin\kappa) \qquad (20)$$

2.3. ACHIEVING PRODUCTIVITY TARGETS WITH THE LEAST POSSIBLE OF WEIGHT

This integral machine and process approach aims at achieving the critical depth of cut shown in (19) with a minimum of material content in the machine. The proposed approach is that, if a specific machine mode does not limit the productivity of the machine, its modal stiffness can be reduced without affecting the productivity and stability of the machining process. And if one specific mode limits the productivity of the machine in one specific milling direction at a specific speed, according to (19) and (20), the productivity of the machine can be maintained or even improved while reducing the mass of the machine by means of the following two complementary approaches:

1) By modifying the real directional factor σ for reducing its absolute value.
2) By reducing in a deliberated manner the modal stiffness associated to that mode –with a subsequent reduction in the mass of involved machine components - and by increasing in parallel the modal damping in the same proportion by means of active damping devices, maintaining thus constant the product $k_{ef}\xi$.

Next, these two complementary approaches for assuring a specific value of productivity with a minimum of machine weight will be analysed.

2.3.1. Modification of the Directional Factor σ for Reducing the Need of Machine Weight for Specific Productivity Levels

The *real directional factor* σ is a dimensionless factor that depends on the relative orientation between the feed velocity of the tool and the modal vector and on the cutting coefficients normalised with respect to the tangential coefficient. For one mode "q" of the machine, according to equations (16) and (18), it takes the following form:

$$\sigma_q = \int_{\phi_0}^{\phi_1} a_{h_q} (\cdot a_{F_t}^q + K_r \cdot a_{F_r}^q + K_a \cdot a_{F_a}^q) d\phi \qquad (21)$$

where a_F^q refers to the projection of a force F on the direction of a mode "q", and a_{h_q} refers to the projection of the modal displacement associated with mode "q" on the direction of the chip thickness h_d. Low absolute values of this factor improve the stability of the process, since the critical depth of cut will tend to increase, as can be deduced from (19).

As it has been shown in Figure 8, this directional factor σ varies in a notable manner with respect to angles β_z and β_{xy} (angles between the mode and the working plane and between the projection of the mode on the working plane and the tool feed direction respectively). As a reference,

Figure 9 shows how this directional factor varies on the working plane for different β_z angles,

Figure 9 shows that directional factors oscillate in a notable manner with respect to the values of β_{xy} angle, and the fluctuation characteristics depend on the value of β_z angle: for β_z= 0°, oscillations are low, for β_z= 45° oscillations are more pronounced, and for β_z= 90° those oscillations disappear, with constant factors.

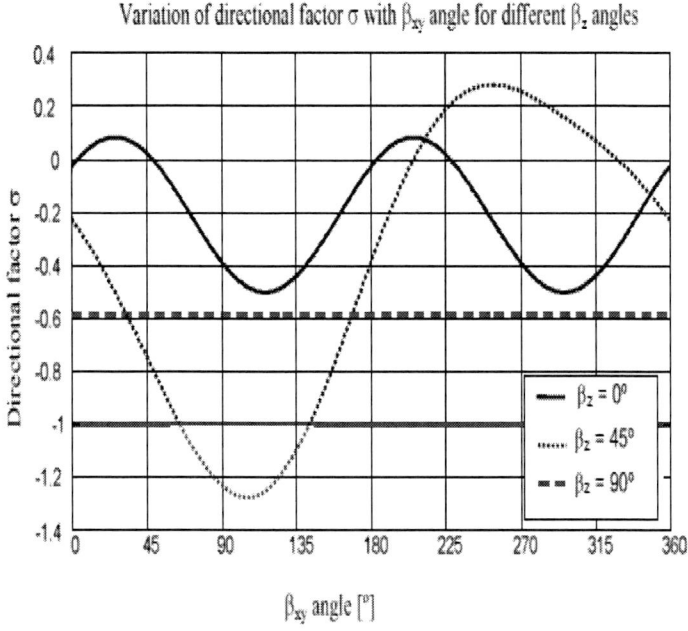

Figure 9. Variation of directional factor σ with respect to βxy angle for different β_z angles.

Figure 10.a shows in turn the worst case of σ from the process stability point of view, that is the maximum absolute value, for different β_z angles (having swept β_{xy} from 0° to 360° for each of the β_z angles). Finally, Figure 10.b shows the average value of absolute values of σ for different β_z angles (having swept β_{xy} from 0° to 360° by steps of 1°).

Figure 10.a shows that in the range of β_z angles from 22° to 72°, there are β_{xy} angles, i.e. machining directions in the $X_{MT}Y_{MT}$ working plane, where the values of the directional factors are above unit in absolute value, |σ|>1, which leads to reduced values of effective stiffness. The same info can be extracted from the 2D map that has been shown in Figure 8 for a specific β_z angle. In calculating average values of directional factors when sweeping the range of machining directions from 0° to 360° with a resolution of one degree, it turns out that for increasing values of β_z angles, the average value of the directional factor, in absolute value, also increases.

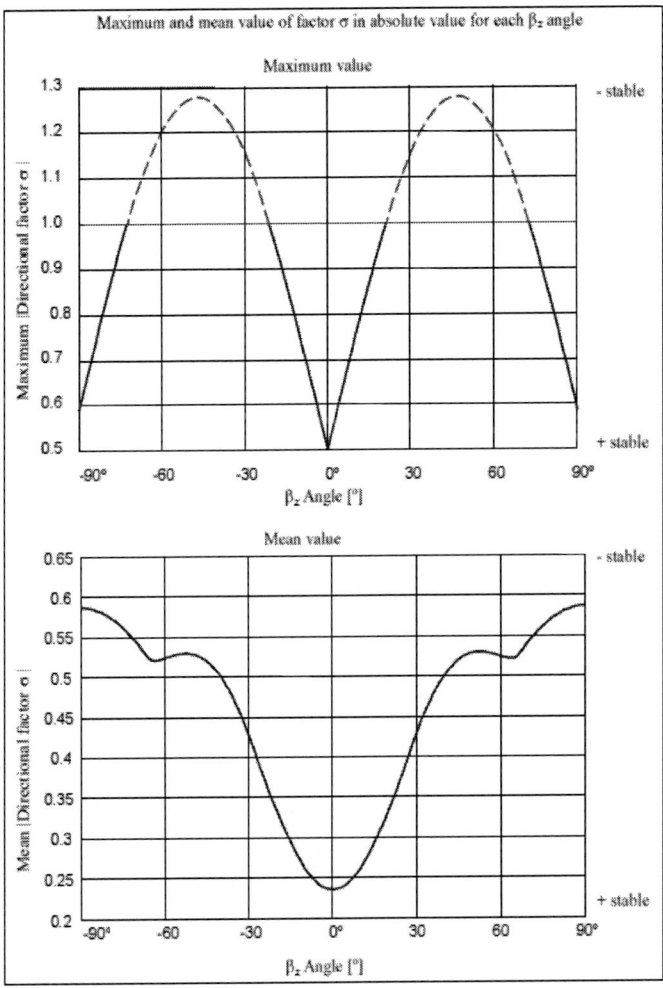

Figure 10. a) Above: Maximum absolute value of directional factor σ for different βz angles. b) Below: Average of absolute values of directional factor σ for different βz angles.

Therefore, it can be concluded that for the cases where there are predominant machining directions, e.g. roughing of large parts in the longitudinal direction, the machine can be designed such that their structural modes and the predominant machining directions do not lead to combinations where directional factors are above unit. For the case that there are not predominant machining directions, the general rule is that for the foreseen dominant modes, the lower the β_z angle the more stable the machining process, as shown in Figure 10.b. That means that for the case that the modal vector is contained in the $X_{MH}Y_{MH}$ working plane, i.e. $\beta_z=0°$, the machining processes will be more stable, or in other terms, there will be margin during the design stage for reducing the machine stiffness -and consequently the material intensity of the machine- for an aimed value of machining productivity.

This approach is not easy to be applied in practice, since mode directions are highly dependant of machine architecture, so that changing the direction of modes is not a trivial issue. One interesting alternative to changing the direction of modes lies in changing the machine table orientation, as shown in Figure 11, because this way the angle between the considered machine mode and the working plane, i.e. the β_z angle, is changed without changing the machine architecture.

As reference, Figure 11 shows a milling machine with an architecture based on a mobile column and an embedded horizontal ram. For that machine architecture, it is quite common that the predominant mode is the bending mode in the vertical direction of the horizontal ram, as shown in Figure 11 above, so that the modal vector associated with that mode will be approximately a vertical vector. On the other hand, it must be noted that the X_{MT} and Y_{MT} axes are contained within the working plane, so that if the working plane is re-oriented, also the directions of the X_{MT} and Y_{MT} axes are modified, so that the vector components of the bending mode have also to be adapted.

In this case, if instead of changing the direction of the predominant mode, the orientation of the $X_{MH}Y_{MH}$ working plane is changed, the angle between the bending mode and the working table will pass from the initial $\beta_z \approx 90°$ to $\beta_z \approx 0°$, i.e. the modal vector associated with the bending mode passes from being normal to the working plane ($\beta_z \approx 90°$) to being contained within the working plane ($\beta_z \approx 0°$).

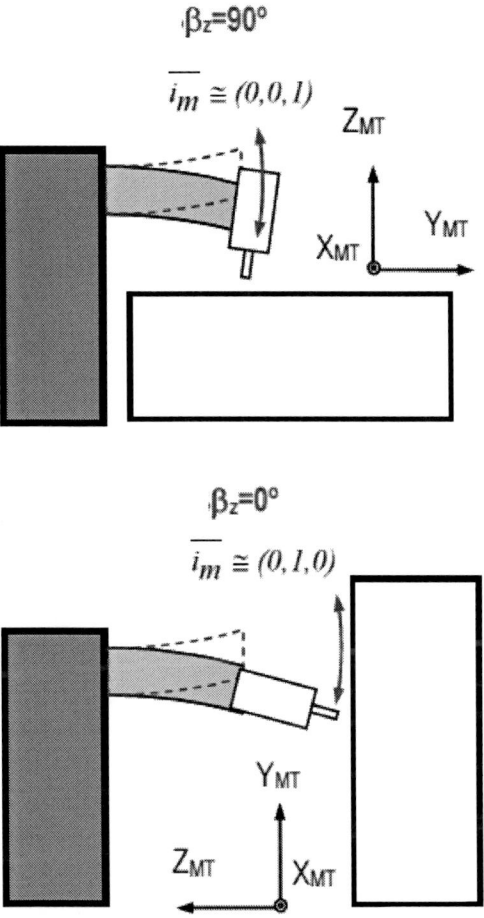

Figure 11. Change in the working plane orientation for modifying the directional factor σ and thus the process productivity. *Above*: Initial machine architecture; *Below*: re-designed machine architecture.

This means that it is possible to increase the stability of a machining process and the productivity of the machine without modifying the mechanical robustness or the material intensity of the machine structure. Therefore, for a given process productivity goal, there is a possibility of maximizing the directional factor σ by adapting the orientation between the mode and the working plane, thus gaining a margin for reducing the dynamic stiffness associated with the predominant modes and consequently the mass associated with the machine without affecting its productivity.

2.3.2. Modification of the Product between Dynamic Stiffness and Damping Coefficient to Reduce the Need of Machine Weight for Certain Productivity

The second approach to improve the productivity of a machine while reducing the mass of the machine is based on the fact that it is feasible to maintain the critical depth of cut of a machining process while decreasing the stiffness and mass associated with the limiting mode of the machine, provided that in parallel its modal damping is increased by a set percentage, so that the product $k \cdot \xi$ of stiffness and damping of the limited mode is maintained, as shown in (20).

Within the approach of increasing the modal damping of a limiting mode as a means for gaining margin for reducing the modal stiffness, there are systems that provide additional damping at a range of frequencies or at a given direction, such as viscoelastic materials, viscous fluids and above all, *Active Damping Devices* (ADDs). A typical ADD consists of a vibration sensor, an inertial actuator and a controller. ADDs are based on the principle that an acceleration of a suspended mass results in a reaction force towards the supporting structure. In order to tune the acceleration an embedded sensor monitors the supporting structure vibration; the sensors readings are sent to an external feedback controller that drives the internal electromagnetic actuator of the ADD. As a result, these devices can damp the vibration modes that they observe in an open-loop transfer function [16].

ADDs add damping to the machine independently of its dynamic properties. The ideal place to locate the ADDs is as close as possible to the tool centre point. Thus, in machines that have a movable ram, a good place to allocate ADDs is at the end of the ram close to the headstock, as illustrated in Figure 12.

One interesting aspect related to the use of these ADDs is the estimation of the additional damping that they can provide, so that during the design stage a parallel reduction of the modal stiffness can be implemented for maintaining the product between the stiffness and the damping with a lower material content.

With the aim of quantifying the increase of damping associated with these ADDs, a couple of ADDs of Micromega Company have been placed on a horizontal ram of a milling machine of Fatronik, so that the horizontal bending mode of the machine can be damped, as shown in Figure 13.

Figure 12. ADD placed on the ram of a milling machine.

Once the ADDs have been placed on the ram of the machine, experimental Frequency Response Functions -FRFs have been measured at the machine TCP, both for the ADD activated and deactivated cases. Figure 13 shows two FRFs in the X_{MT} direction, the first FRF with the ADD activated and the second FRF with the ADD deactivated.

Figure 13 is of great interest, as it shows damping of 60% of the dynamic flexibility associated with the bending mode of the machine, from the initial dynamic flexibility of 4.3e-7 m/N at 32.5 Hz to the final value of 1.7e-7 m/N at 32 Hz. This damping of the mode is equivalent to an *increase of 253% of the relative damping coefficient* ξ associated with that mode in relation to the initial coefficient, considering the system as of one degree of freedom, where the dynamic displacement D_{dyn} is related to the static displacement D_{st} by the following expression:

$$D_{dyn} \approx D_{st} \cdot (1/2\xi). \tag{22}$$

This means that the critical depth of cut for machining processes in which the limiting mode with regard to the appearance of self-induced vibrations is that damped mode, will be around 253% higher with that ADD than the achievable depth of cut without that ADD.

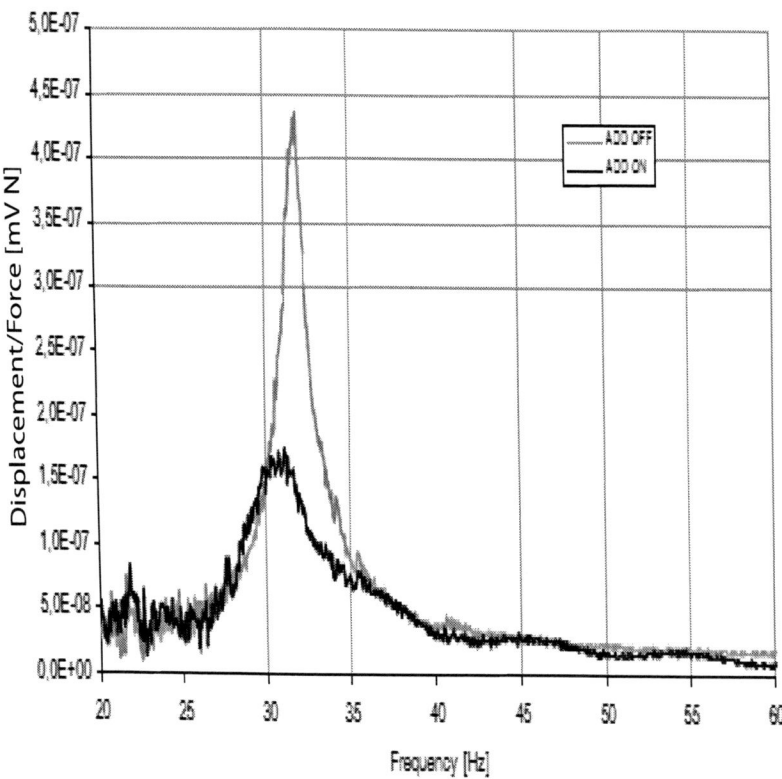

Figure 13. FRF in X_{MT} direction of a horizontal milling machine with and without an ADD.

Taking into account that damping coefficient ξ has increased by 253%, it is possible to maintain the product between stiffness and damping coefficient associated with that mode *by decreasing the effective stiffness by 60%*. Thus, a remarkable reduction can be achieved in the material intensity of the machine without affecting its productivity.

Chapter 3

APPLICATION OF THE APPROACH ON AN ACTUAL INDUSTRIAL CASE

This methodology aimed at integrating productivity and eco-efficiency in milling machines has been applied on an actual industrial case, in concrete on the re-conception of an actual milling machine of a machine tool builder. The selected milling machine has a fixed-table, with a moving column in the longitudinal direction X_{MH}, built-in vertical slide inside the column in the vertical direction Z_{MH} and a built-in horizontal ram inside the slide for the transversal direction Y_{MH}. This machine architecture is commonly known as *box-in-box* type architecture, and is an architecture that is geared towards the general mechanics sector.

The decision for applying the methodology to this family of milling machines has been based on its high ratio between work volume and machine volume, its versatility and, above all, its serial architecture concerning its drives, in the sense that any reduction in weight in an internal axis means a reduction in the weight involved in the external axes. Figure 14 shows an scheme of this machine architecture.

Equipped with a universal milling head, the selected machine reaches feed speeds of 30 m/min and accelerations of 1.2 m/sec ^{2}. Its current column has an approximate height of 3,500 mm and weighs 6.500 kg. The vertical slide contains a section of approximately 1,000 x 1,000 mm^2 and weighs 1,400 kg. The horizontal ram, for its part, contains a section of approximately 500 x 450 mm^2 and weighs 1,450 kg. All these structural elements are made using welded S275 JR steel.

With regard to the transmission of the milling machine, the axis drive Y in the feed direction of the horizontal ram comprises a spindle of 50 mm in diameter, using a pulley transmission and an asynchronous servo-motor of

10 kW rated output. Axis Z, for its part, comprises a spindle of 63 mm in diameter, with pulley transmission and an asynchronous servo-motor of 10 kW rated output.

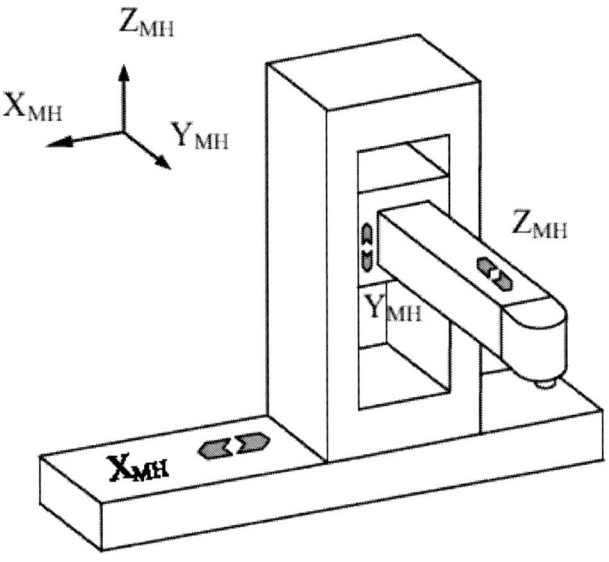

Figure 14. *Box-in-box* architecture of the milling machine to be re-designed.

3.1. ANALYTICAL AND EXPERIMENTAL MODAL STUDY OF THE CURRENT MILLING MACHINE

As a starting point for the re-conception of this actual milling machine, an FEM model has been developed, which has enabled the relevant machine modes to be obtained:

Table 2. Actual values for damping coefficients ξ

Mode #	Actual value for eigen-frequency	Actual value for damping coefficient ξ
1	13.5 Hz.	1.66 %
2	16.2 Hz.	2.32 %
3	28.5 Hz.	2.66 %
4	36.6 Hz.	3.11 %

Application of the Approach on an Actual Industrial Case 29

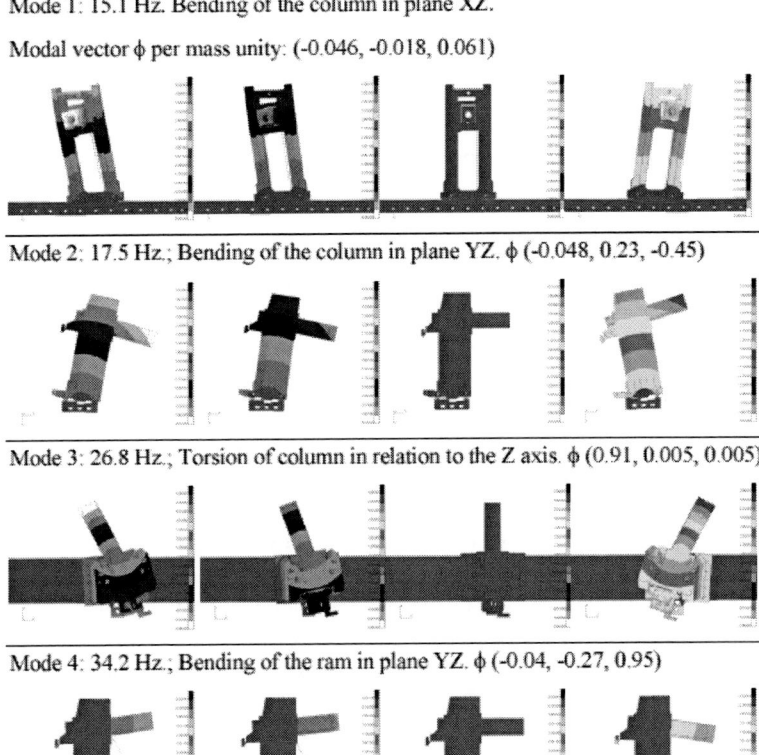

Mode 1: 15.1 Hz. Bending of the column in plane XZ.
Modal vector φ per mass unity: (-0.046, -0.018, 0.061)

Mode 2: 17.5 Hz.; Bending of the column in plane YZ. φ (-0.048, 0.23, -0.45)

Mode 3: 26.8 Hz.; Torsion of column in relation to the Z axis. φ (0.91, 0.005, 0.005)

Mode 4: 34.2 Hz.; Bending of the ram in plane YZ. φ (-0.04, -0.27, 0.95)

Figure 15. Analytical modal study of the original milling machine to be re-designed.

This analytical study has been complemented by an experimental modal analysis, which has enabled the real values of the machine's frequencies to be measured and, above all, the real values of the modal damping coefficients.

Next, additional analytical calculations have been conducted on the FEM model of that actual milling machine, with the goal of having as much information as possible about the limiting modes of the machine as well as about the directional factors, within an aim of removing as much weight as possible from the structural components of the machine without affecting its final productivity in terms of MRR.

Thus, Figure 16 shows the values of modal stiffness, effective stiffness and the quotient between the effective stiffness and the maximum absolute value of the real directional factor σ associated with that mode in its working plane.

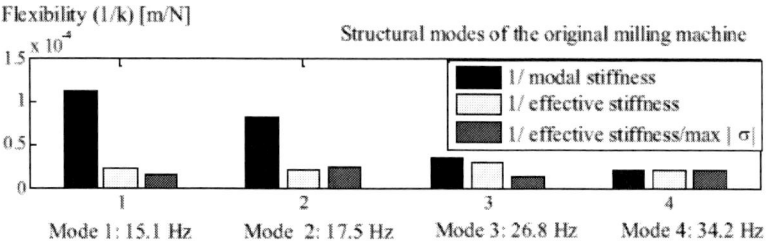

Figure 16. Additional analytical calculations on the FEM model of the original milling machine.

As it can be seen in Figure 16, the directional factor σ can increase the flexibility associated with a certain machine mode (such as for the case of mode 2), whereas in other cases, it reduces the flexibility associated with the mode (such as for the cases of modes 1, 3 and 4).

On the other hand, based on the calculations that have been shown in Figure 16, the dynamic stiffness associated to these four modes have been calculated in the three machine axes X_{MT}, Y_{MT} and Z_{MT}, showing the results in Figure 17.

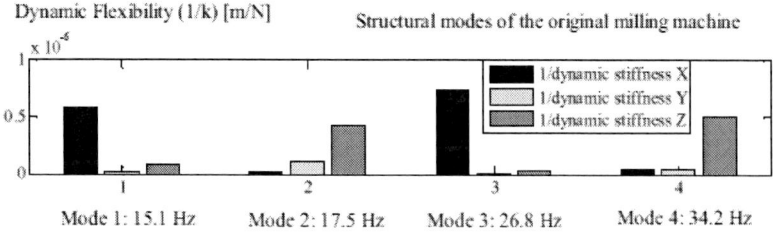

Figure 17. Dynamic stiffness associated to the original milling machine modes in Cartesian axes.

As it can be seen from Figures 16 and 17, there is not a direct relation among the modes with highest effective stiffness divided by the directional factor σ on the one hand and the modes with highest dynamic flexibility on the other hand. This aspect is of great interest for the re-design process for milling machines, since allows joining weight reduction strategies (linked to the dynamic stiffness of a machine) and productivity increase strategies (linked to the effective stiffness divided by the directional coefficient σ).

Next, the stability lobe diagrams have been calculated for these FEM model of the machine and for a typical milling operation, with the aim of identifying *which is the mode or modes that currently limits the productivity of the machine* in given machine positions and machining directions.

Thus, for calculating the stability lobes, the following machining process has been considered: an AISI 1045 steel rough milling operation with a 125 mm plate, 9 flutes at a 45° angle and with 80% radial immersion in concordance, and a working range of 400 to 600 rpm for the tool.

Figure 18 shows a polar diagram that shows the critical depths of cut that have been obtained in this machine for the machining process mentioned above and with the horizontal ram in an outer and lower position. In that polar plot, 0° coincides with the $+X_{MT}$ direction of the machine, and the 360° cover the different machining directions within the $X_{MT}Y_{MT}$ working plane.

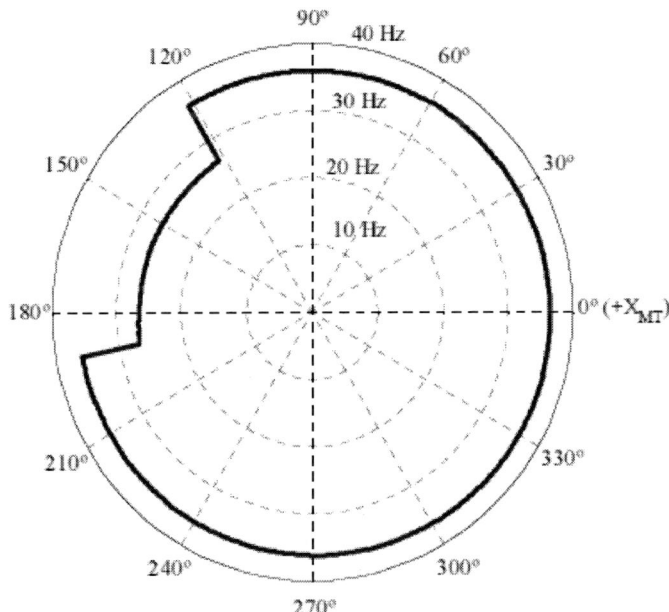

Figure 18. Polar diagram of structural chatter frequencies in the original milling machine. Spindle speed: 400=600 rpm.

As it can be seen in Figure 18, for the great majority of machining directions, the chatter frequency is in the range of 35 Hz (roughly speaking, in the first, the third and the fourth quadrants of the working plane) and in the range of 25 Hz (roughly speaking, in the second quadrant of the working plane). From a identification process among the chatter frequencies and the structural frequencies of the machine listed in Figure 15, it can be concluded that the chatter frequencies of this machine/process combination are associated with mode 3 involving vertical torsion of the column (at 26.8 Hz.) and mode 4 involving flexion of the ram and the frame on the vertical plane (at 34.2 Hz.). The fact that in some cases the chatter frequencies are below the natural frequencies and in other cases the chatter frequencies are above the natural frequencies is associated with the sign of the directional factor σ, as it can be concluded from Figure 7.

From these data, some conclusions of interest can be drawn:

I. If the mass and rigidity associated with these two limiting modes 3 and 4 is reduced, combined with a parallel increase in the same proportion of the damping associated with such modes, the productivity of the machine will be able to be maintained with using less material resources and with a reduction in energy consumption in the phase of use.

II. If the mass and rigidity associated with non-limiting modes of the machine such as modes 1 and 2 is reduced by continuously checking to ensure these changes do not alter the non-limiting nature of such modes, the productivity of the machine will be able to be maintained with using less material resources and with a reduction in energy consumption in the phase of use.

3.2. RE-DESIGN OF THE HORIZONTAL MILLING MACHINE

From the analytical and experimental study carried out on the horizontal milling machine, the following changes in mechanical design have been implemented in order to reduce the weight of the machine while at the same time maintaining and even increasing its productivity:

1) Lightening of the ram by using light materials and a high degree of internal damping: a ram has been designed based on steel sandwich and aluminium foam elements that reduces the total moving mass by 15% and increases internal damping by 250% in relation to a conventional steel ram, thus reducing static rigidity in directions X

and Z by 10%. The ram guide systems have also been re-conceived in order to increase the damping associated with the ram flexion mode without this meaning an increase in the mass of the machine.
2) Lightening of the milling head: by replacing steel with cast aluminium, the weight is reduced by 27%, decreasing from the initial 550 kg to 400 kg without this affecting the productivity of the machine.
3) Reduction in conductive inertia: by reducing the inertia associated with the moving components and optimizing the dynamic behaviour of the servo-drives, the total inertia reflected in the drive shaft has been able to be reduced by 50%, thus using servo-motors with less inertia in the servo-axes Y and Z.
4) Introduction of active damping systems: in order to improve the damping associated with the torsion mode of the column without modifying the column itself, an active vibration control system has been incorporated into the horizontal ram based on an inertial actuator whose direction of application of force is the X axis.

The above changes have been introduced into the MEF model of the machine, which has enabled the effects of such changes to be virtually validated both with regard to the weight of the moving elements and the productivity of the machine - measured in terms of the amount of chip that the machine is capable of evacuating without there being instability.

3.3. MANUFACTURE OF A PROTOTYPE OF A PRODUCTIVE AND ECO-EFFICIENT MILLING MACHINE

Once the original horizontal milling machine has been re-designed in accordance with the criteria indicated in the previous subchapter, it can then be manufactured and assembled so as to be able to experimentally check the reductions in weight and increases in productivity when incorporating the different re-design techniques, in addition to the active vibration control system.

The mechanical and dynamic results that have been obtained regarding this prototype of productive and light milling machine, prior to carrying out consumption and productivity measuring tests, are as follows:

Table 3. Comparison of dynamic and mechatronic properties between the original milling machine and the Prototype of light and productive machine; data from Y axis

Y axis of the machine (longitudinal direction of the ram)			
Concept	Original machine	Prototype	Comparison
Maximum acceleration / continuous rating	1.3 / 3.1 m/s^2	1.3 /2.9 m/s^2	The dynamic functional nature of the machine is maintained
Moving mass	2,500 kg.	2,100 kg	16% reduction in the moving mass
Mass reflected in the motor	13,650 kg	7,085 kg	48% reduction in the mass reflected in the motor
Mechanical rigidity of the transmission	328 N/μm	219 N/μm	33% increase in the mechanical rigidity
Natural frequency in the drive	29 Hz	35 Hz	20% increase in the natural frequency
Bandwidth of the drive	6 Hz	7 Hz	21% increase in the mechatronic robustness
Output of the motor	10 kW	6 kW	40% reduction in the motor output

Table 4. Comparison of dynamic and mechatronic properties between the original milling machine and the Prototype of light and productive machine; data from Z axis

Z axis of the machine (vertical direction of the milling machine)			
Concept	Original machine	Prototype	Comparison
Maximum acceleration / continuous rating	1.3 / 2.1 m/s^2	1.3 /1.9 m/s^2	The dynamic functional nature of the machine is maintained
Moving mass	4,500 kg.	3,700 kg	18% reduction in the moving mass
Mass reflected in the motor	15,415 kg	8,965 kg	42% reduction in the mass reflected in the motor

Application of the Approach on an Actual Industrial Case

Concept	Z axis of the machine (vertical direction of the milling machine)		
	Original machine	Prototype	Comparison
Mechanical rigidity of the transmission	430 N/μm	300 N/μm	30% reduction in the mechanical rigidity
Natural frequency in the drive	26 Hz	30 Hz	15% reduction of the natural frequency
Band width of the drive	5 Hz	6 Hz	17% increase in the mechatronic robustness
Output of the motor	10 kW	6 kW	40% reduction in the motor output

Below are shown the *Frequency Response Functions* (FRF) which have been measured in the prototype of a productive and light-weight machine, in addition to their comparison with the original milling machine with similar functions. As a starting point, Figure 19 shows a comparison between FRFs in direction X of the machine.

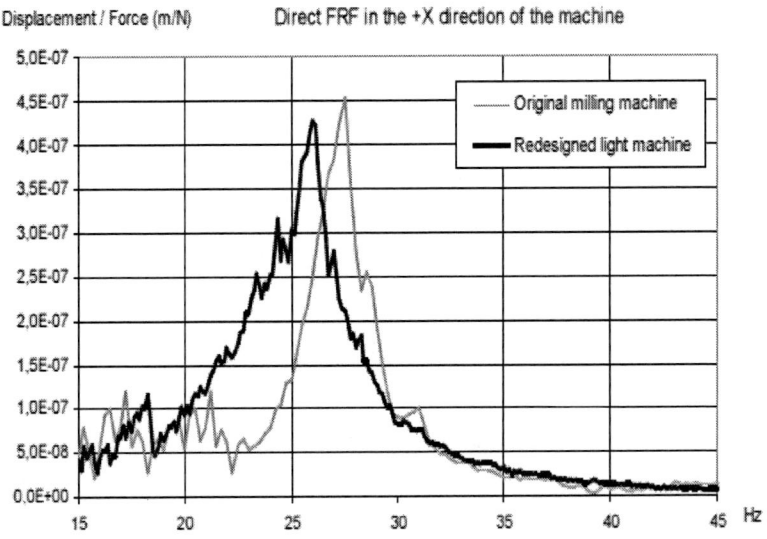

Figure 19. Comparison between direct FRFs in direction X: original milling machine vs. Prototype of light-weight machine.

By comparing both FRFs, it can be seen how the structural mode of the initial machine, associated with torsion of the column, has been transformed into a mode with more passive damping and with greater dynamic flexibility which has remained practically invariable. Both effects are positive from the eco-efficiency point of view, as on the one hand the Prototype of light and productive machine has less mass (the ram weighs 15% less) which yet does not jeopardize productivity due to an increase in damping. On the other hand, it should be taken into account that no active vibration control system has been included in these FRFs – an aspect that will result in an additional increase in productivity at the expense of a minimal increase in energy consumption.

Furthermore, Figure 20 shows another comparison of direct FRFs between the redesigned machine and the original milling machine, in this case for direction Y of the machine (horizontal feed of the ram).

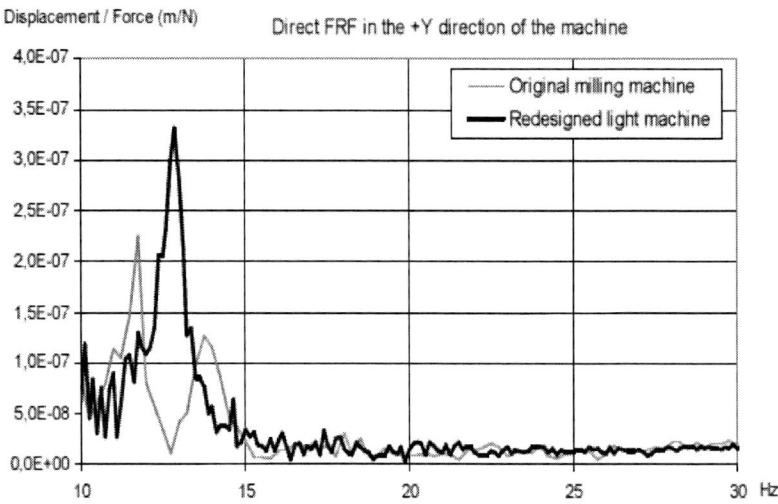

Figure 20. Comparison between direct FRFs in direction Y: original milling machine vs. Prototype of light-weight machine.

In this case, the Prototype evidences less dynamic rigidity than in the initial machine. This does not constitute any problem: on the contrary, it has been the result of a deliberate strategy, as the stability lobe diagrams indicated that the machine modes on the Y axis did no limit the productivity of the machine. This has enabled the ram to be re-designed with less dynamic rigidity (and therefore with less weight) without affecting the productivity of the machine. Lastly, Figure 21 shows the comparison of FRFs in this case for direction Z (vertical movement of the frame).

its moving components, the productivity of the machine can be increased while at the same time consuming less energy while it carries out machining operations.

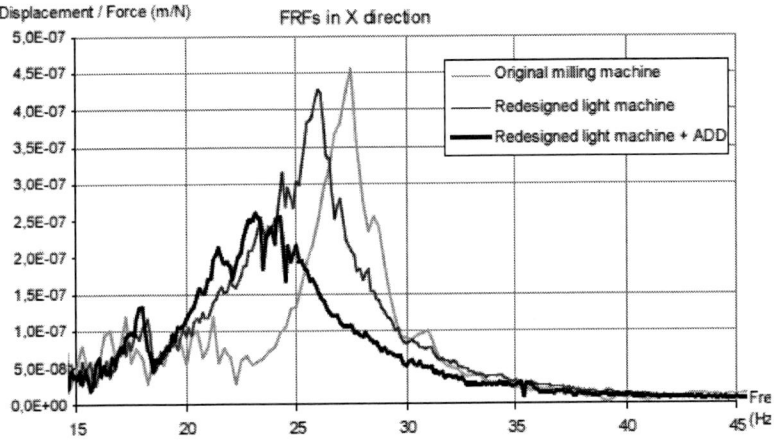

Figure 23. Direct FRFs in X_{MT} direction: Original milling machine vs. Prototype of light-weight machine, with and without ADD.

Once the new FRFs have been measured at the TCP of the Prototype, the stability lobe diagrams have then been calculated for a milling process with a tool of 125 mm in diameter and nine teeth, machined in different directions within the XY work plane of the machine. The diagrams have been calculated initially with the active damper deactivated, and subsequently with the damper activated.

Figure 24 shows eight different stability lobe diagrams, which correspond to gradual variations in the machining direction at 45°. The maximum depth of cut is shown for each machining direction for different rotational speeds of the tool, and also for the cases of the activated and deactivated vibration control system.

On the other hand, for the machining direction at 180° (direction X- of the machine), and focusing on the range of working speeds of the tool that is being taken into consideration – i.e., from 300 to 700 rpm – the graphs are obtained which are shown in Figure 25. For this case, the lobe diagram on the left shows increases higher than 100% in terms of maximum depth of cut values – an improvement that is associated with the increase in modal damping in structural modes that limit the productivity of the machine.

In the diagram on the right can be also seen how the active vibration control associated with the structural mode that limits the productivity of the machine, which is in a lower frequency mode, gives rise to a jump in the vibration frequency. This means that the initial bending mode (i.e. mode 4) stops being limiting and a mode of a higher frequency becomes limiting, with significantly greater modal rigidity.

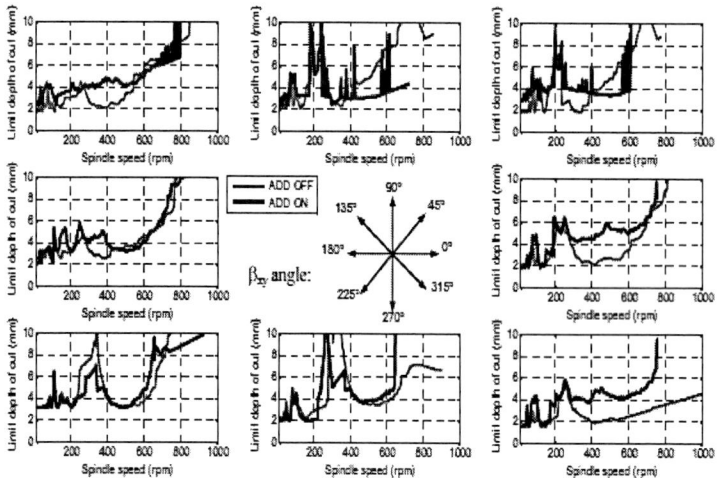

Figure 24. Stability lobe diagrams of the Prototype with and without active vibration control for different machining directions (β_{xy} angle) in the working plane $X_{MT}Y_{MT}$.

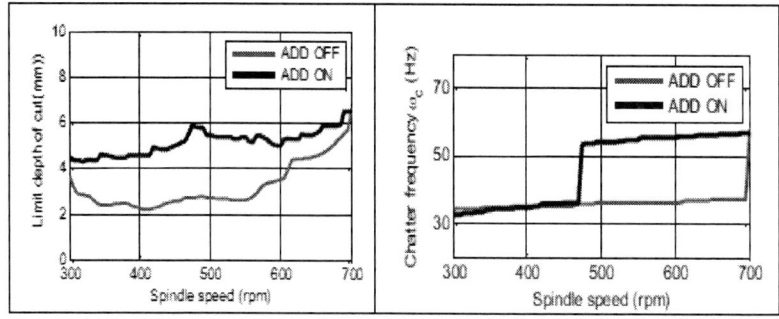

Figure 25. Zoom of the active damping in direction X=180° (X-) of the Prototype; rotational speeds of 300 – 700 rpm.

Chapter 4

VALIDATION OF THE PROTOTYPE OF PRODUCTIVE AND ECO-EFFICIENT MILLING MACHINE

4.1. TESTS FOR MEASURING PRODUCTIVITY: MATERIAL REMOVAL RATE MRR

The active damping effect has proved especially noticeable in the direction of actuation, which in the Prototype ram has been direction ± X. The increase in modal damping in the bandwidth of actuation of the Prototype has enabled the dynamic flexibility associated with the milling machine to be reduced and thus obtain an increase in the critical depths of cut.

Figure 26 shows the effect active damping has meant in the different machining directions within the $X_{MT}Y_{MT}$ work plane of the Prototype. This Figure shows a polar diagram that relates machining directions and maximum axial depths of cut that prevent the appearance of instability in the machining process. 0° coincides with the $+X_{MT}$ direction of the machine.

Figure 26 shows how in machining direction $+X_{MT}$ (0° angle in the polar diagram), the axial critical depth of cut of the Prototype has been increased by over 100% when activating the active control of vibrations. This is equivalent to an increase in the productivity of the machine in MRR to the same extent – that is, by 100% - when the increase in energy consumption is lower, as the inertial actuator of which the vibration control system consists operates at a maximum output of 200 watts, i.e. around 0.5% of the total output of the Prototype, and the mobile structural elements involved in the $+X_{MT}$ motion weigh approximately 20% less.

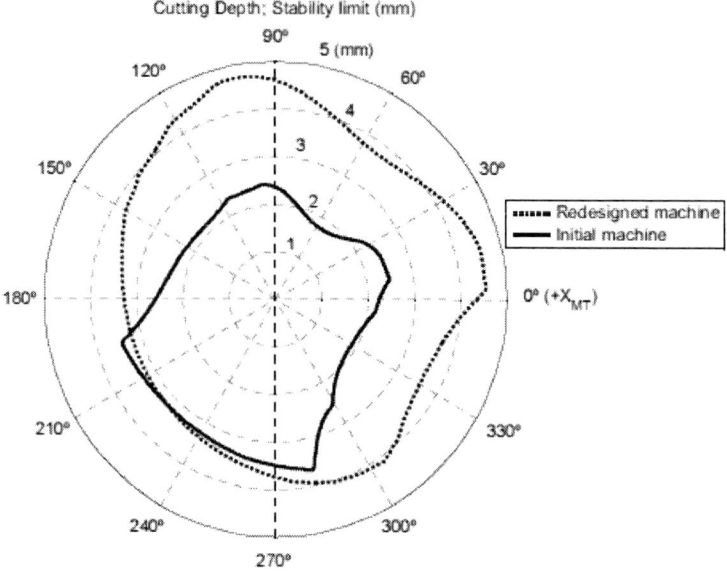

Figure 26. Productivity of the Prototype of light and productive machine in different machining directions.

This increase in productivity is not confined to the direction of the actuator. Figure 26 indicates that in machining direction Y+ (90° in the polar diagram), the maximum critical depth of cut is also increased by 100%, with the vibration control system activated – in other words, in machining direction Y+, the Prototype has obtained a further 100% increase in the productivity of the machining processes.

If it is taken into account that the Prototype has obtained an increase in productivity of 100% in directions +X y +Y, and that this increase in productivity has been obtained in a milling machine that consumes less energy than a conventional milling machine of the same functional nature, the conclusion can be drawn that the Prototype combines an increase in productivity and an increase in energy efficiency in such a way that the Prototype validates a new concept in the machine tool sector: eco-productivity.

In the following section the reduction in energy consumption of the Prototype will be stressed in more detail and in a more quantified way.

4.2. TESTS TO MEASURE ENERGY CONSUMPTION

4.2.1. No-Load Tests

The lightened features of the Prototype, with reduction in weight of over 40% in the inertias reflected in drives Y and Z, mean that its servo-mechanisms reach high feed speeds and high rates of acceleration while at the same time consuming less energy. Expressed in a quantified way, Figure 27 shows a comparison between the energy consumption associated with the redesigned light-weight machine and a conventional milling machine of the same dimensions and functions, when both milling machines are subjected to the same positioning references and in both cases acceleration is the maximum that can be obtained by their servo-mechanisms.

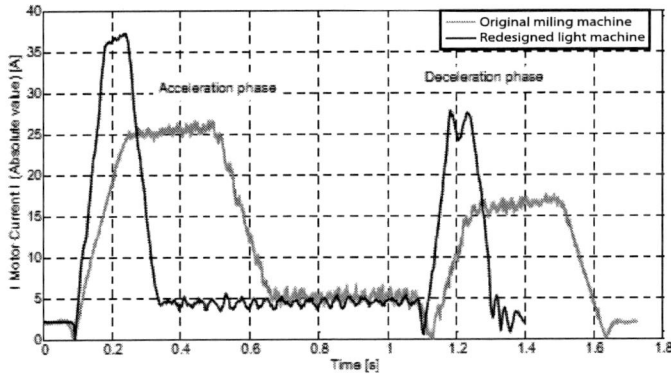

Figure 27. Comparison of energy consumption between axis-Y motors in the Prototype of light and productive milling machine and the original milling machine.

Specifically, Figure 27 shows the temporary movement profiles that are associated with the positioning of the tool at 30 m/min and at the maximum acceleration the servo-mechanism is able to reach. The continuous line shows this positioning process for the original milling machine, whereas the dotted line shows the same case of positioning, but in this case for the Prototype.

Integration over time of the power consumed by the motors of the Y servo-mechanism in both cases indicates that the conventional milling machine consumes 31.9 joules per each positioning cycle, whereas commanding the same profile of movement, the Prototype consumes 21.04 joules during the same cycle. This means that the Prototype of eco-efficient milling machine maintains and even increases productivity in terms of *material removal rate* while at the same time consuming 35% less energy during no-load movements of the tool.

4.2.2. Machining Tests

Below are shown some machining tests that have been carried out with the Prototype of a productive and eco-efficient milling machine. Measurements have been taken for certain AISIS 1045 steel roughing operations with a plate of 125 mm, 9 flutes at a 45° angle and with 80% radial immersion in concordance, rotation of the milling head at 400 rpm and 720 mm/min feed speed – that is, with a feed per tooth of 0.2 mm. The tests have been carried out for different depths of cut and with the vibration damping system activated and deactivated.

Thus, Figure 28 shows the active power consumed by the milling head in certain roughing operations in which the vibration damping system has been activated in some sections.

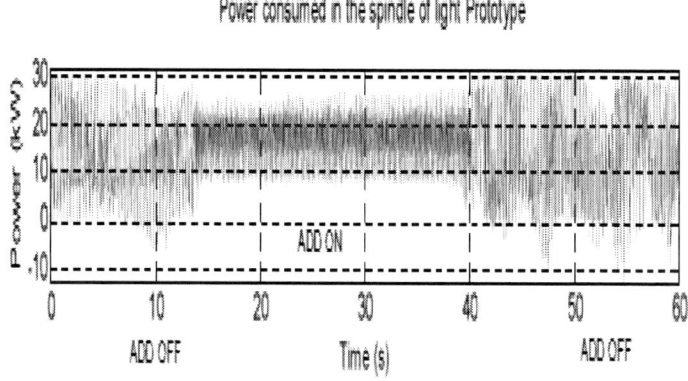

Figure 28. Active power consumption in the head of the Prototype, with and without ADD.

For its part, Figure 29 shows the data from Figure 28 with greater resolution on the axis of abscissa (time axis). In this figure it can be seen that active damping of vibrations reduces both the peaks and the average consumption of active power in the milling head.

Validation of the Prototype of Productive and Eco-Efficient Milling ... 45

By integrating the above power consumption levels in the domain of time, a comparison can be made between the energy consumption in kW·h at intervals of activated and deactivated active damping of vibrations. For machining intervals of 10 seconds, energy consumption with control deactivated has been 0.043 Kwh, whereas with the active vibration damping system on, consumption for the same interval has been 0.040 Kwh - i.e. a reduction of 7% has been obtained in the consumption of electrical energy.

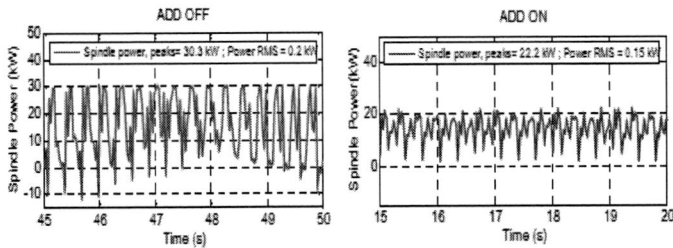

Figure 29. Power consumed by the milling head with ADD activated/deactivated.

On the other hand, active vibration control also reduces the energy consumed by the servo-drive in the feed direction. Thus, Figure 30 shows a comparison between the instantaneous current consumed by the servo-drive motor of the X axis with the vibration damping system activated and deactivated.

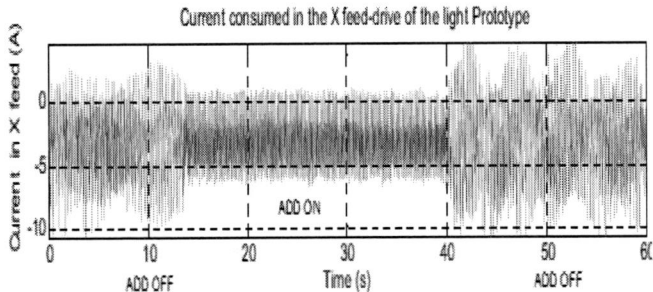

Figure 30. Comparison of consumption of current in drive X with ADD activated/deactivated.

Expressed in a quantified way, the reduction in consumption peaks has been 37%, whereas the average power consumed by the drive has been reduced by 17%.

Chapter 5

CONCLUSION

This book has introduced a design methodology aimed at conceiving milling machines in which high productivity is combined with reduced environmental impact – an aspect that is related to a large extent with the consumption of material resources in the construction phase and energy consumption in the phase of use. The approach has been based on breaking the link between a stable machine-part interaction during the machining process and the need to obtain high levels of mechanical rigidity by means of structural components of great weight.

The construction of a Prototype of eco-efficient and productive milling machine, in which reductions of up to 50% have been obtained in the inertias reflected in the drive shafts in relation to the conventional milling machine, has enabled reductions of 35% to be obtained in the energy consumed in non-load movements in relation to the milling machine referred to above. Parallel to this, increases in the *material removal rate* of 100% have been obtained in some machining directions. On the other hand, the integration of an active vibration control system has enabled reductions in energy consumption to be obtained both in the milling head itself and in the feed servo-drives in terms of the energy consumption of the Prototype, without vibration damping.

All this leads us to draw the conclusion that a multidisciplinary approach aimed at maximizing the ratio between the mass involved in a machine tool and the productivity level reached with that machine has a beneficial effect both on the machine supplier and the its user as, on the one hand, manufacturing and machine purchase costs are reduced and, on the other, operating costs are reduced. Indeed, this mutual benefit, combined with a modular conception of the machines that enables light and productive machines to be put together swiftly, may in turn enable new business models to be established in the manufacturing sector among suppliers and users of machine tools, such as sharing benefits, risks and costs associated with manufacturing processes.

REFERENCES

[1] Dietmair A. et al. (2010), Lifecycle Impact Reduction and Energy Savings through Lightweight Eco-Design of Machine Tools. 17th CIRP International Conference on LCE, Anhui, China.

[2] Henninger C., Eberhard P. (2007), An Investigation of Pose-Dependent Regenerative Chatter for a Parallel Kinematic Milling Machine. Proceedings of the IFToMM 12th World Congress in Mechanism and Machine Science, Besançon, France.

[3] Bravo U. (2007), Un procedimiento para la predicción de la estabilidad dinámica en el mecanizado a alta velocidad de paredes delgadas (in Spanish). Doctoral Thesis in the University of the Basque Country UPV/EHU.

[4] Dequidt A. et al. (2000), Mechanical pre-design of high performance motion servomechanisms. *Mechanism and Machine Theory* 35 pp. 1047-1063.

[5] López de Lacalle, L.N.; Lamikiz, A. (Eds.) (2009), Machine Tools for High Performance Machining, pp. 1-44.

[6] Zulaika J. et al. (2010), Eco-efficient and highly productive production machines by means of an holistic Eco-Design approach. Proceedings of the E/E 3rd International Conference on Eco-Efficiency. Egmond aan Zee, Netherlands.

[7] Zulaika J. Campa, F.J. (2008), Optimización de los parámetros de diseño de una máquina-herramienta en base a criterios de productividad y ecoeficiencia (in Spanish), XVII Congress on Machine Tools and Manufacturing Technologies, San Sebastián, Spain.

[8] Altintas, Y. (2001), Analytical Prediction of Three Dimensional Chatter Stability in Milling, JSME Int. J. Series C: *Mechanical Systems, Machine Elements and Manufacturing,* 44, n.3.

[9] Zatarain, M., Muñoa, J., Peigné, G., Insperger, T. (2006), Analysis of the Influence of . Mill Helix Angle on Chatter Stability, *Annals of CIRP,* Vol. 55, No. 1, pp 365-368.

[10] Zulaika J. et al. (2010), Highly productive machining processes and eco-efficient machine tools by means of an integrated machine+process approach, Proceeding of the 4th International Conference of High Speed Machining ICHMS2010. Guangzhou (China).

[11] Altintas, Y., Budak, E. (1995), Analytical Prediction of Stability Lobes in Milling, Annals of the CIRP, 44/ 1: 357-362.

[12] Zatarain, M., Insperger, T., Peigne, G., Villasante, C., Muñoa, J. (2007), Analysis of Directional Factors in Milling: Importance of Multifrequency Calculation and of the Inclusion of the Effect of the Helix Angle, 6th Int. Conf. on High Speed Machining, San Sebastian, Spain.

[13] Muñoa, J., et al (2005), Optimization of Hard Material Roughing by Means of a Stability Model, 8th CIRP Int. Workshop: Modelling of Machining Operations, Chemnitz.

[14] Zulaika J. et al (2009), Stability Lobe Diagrams for the Redesign of a Machine-Tool based on Ecoefficiency Criteria, 12th CIRP Conf. on Model. of Mach Operations, San Sebastian.

[15] Altintas, Y., (2000), Manufacturing Automation: Metal Cutting Mechanics, Machine Tool Vibrations, and CNC Design; Cambridge University Press.

[16] Ganguli, A. Deraemaeker, A. Preumont, A. (2007), Regenerative chatter reduction by active damping control, *Journal of sound and vibration,* vol. 300, pp. 847-862.

INDEX

A

Active Damping Device, 24, 38
actuation, 41
actuators, 2, 38
aluminium, 32, 33
asymmetry, 15

B

bandwidth, 41
base, 49
bending, 22, 24, 26, 40
beneficial effect, 47
benefits, 48
business model, 48

C

China, 49, 50
clarity, 9
collisions, 37
conception, 27, 28, 47
concordance, 31, 44
Congress, 49
construction, 47
consumption, 33, 37, 44, 45, 47
cutting force, 9, 13

D

damping, 13, 16, 18, 19, 24, 26, 28, 29, 32, 33, 35, 37, 38, 39, 40, 41, 44, 45, 47, 50
Damping Coefficient, 24
deformation, 6, 11
dematerialization, 3
depth, 9, 12, 15, 17, 18, 19, 24, 26, 38, 39, 41, 42
displacement, 11, 12, 13, 19, 26

E

electromagnetic, 24
end-users, 5
energy, 1, 2, 3, 32, 36, 37, 38, 41, 42, 43, 44, 45, 47
energy consumption, 32, 36, 37, 38, 41, 42, 43, 44, 47
energy efficiency, 42
environmental impact, 1, 2, 47
excitation, 7, 8, 13, 38

F

FEM, 28, 29, 30, 31
flexibility, 26, 30, 35, 37, 38, 41

force, 6, 13, 19, 24, 33
France, 49
freedom, 15, 16, 17, 18, 26
Frequency Response Function, 25, 35, 38

G

geometry, 7
Guangzhou, 50

H

height, 27
housing, 37

I

ideal, 24
identification, 32
immersion, 7, 31, 44
incidence, 6
inertia, 33
integration, 47

L

lead, 5, 9, 17, 22
Life Cycle Assessment, 1, 2
light, 32, 33, 34, 35, 36, 37, 38, 39, 42, 43, 47

M

majority, 1, 6, 31
manufacturing, 47
mass, 8, 11, 17, 19, 23, 24, 32, 34, 36, 37, 47
Material Removal Rate, 1, 41
material resources, 2, 32, 47
materials, 7, 8, 24, 32
matrix, 11, 12, 13, 14, 15

measurements, 6
mechanical performances, 3
methodology, 1, 27, 47
modelling, 8
models, 18
modulus, 10

N

naming, 11, 15
Netherlands, 49

O

operating costs, 47
operations, 5, 6, 39, 44

P

parallel, 19, 24, 32
polar, 31, 41, 42
positive feedback, 13
prototype, 33, 35

R

regeneration, 2, 6
Regenerative Chatter, 49
reliability, 6
requirements, 3, 5, 6, 8
resolution, 20, 44
resources, 3
response, 38
risks, 48
rules, 1

S

sensors, 24
showing, 30
solution, 8, 13, 15
Spain, 49, 50

specifications, 5
spindle, 7, 15, 27
stability, vii, 8, 15, 18, 19, 20, 23, 31, 36, 39
Stability Model, 8, 50
steel, 7, 8, 27, 31, 32, 33, 44
structure, 3, 7, 8, 23, 24
suppliers, 47, 48

T

techniques, 33
teeth, 39
tooth, 13, 44
torsion, 32, 33, 35, 38
transformation, 11

transmission, 27, 34, 35

V

variations, 39
vector, 8, 9, 10, 11, 13, 16, 17, 19, 22
velocity, 1, 19
versatility, 27
vibration, 2, 24, 33, 36, 38, 39, 40, 41, 42, 44, 45, 47, 50

W

weight reduction, 30